The *Incredible* Record-Setting Deep-Sea Dive *of the* Bathysphere

Brad Matsen

Enslow Publishers, Inc.

40 Industrial Road   PO Box 38
Box 398   Aldershot
Berkeley Heights, NJ 07922   Hants GU12 6BP
USA   UK

http://www.enslow.com

For Milo

Library of Congress Cataloging-in-Publication Data

Matsen, Bradford.
  The incredible record-setting deep-sea dive of the bathysphere / Brad Matsen.
    p. cm. — (Incredible deep-sea adventures)
    Summary: Describes the 1934 dive of a bathysphere, or "sphere of the deep,"
in which two explorers, William Beebe and Otis Barton, set the world depth record
and saw mysterious creatures of the deep ocean.
    Includes bibliographical references and index.
    ISBN 0-7660-2188-2 (hardcover)
    1. Underwater exploration—Juvenile literature. 2. Hydrothermal vents—
Juvenile literature. 3. Beebe, William, 1877–1962—Juvenile literature. 4. Barton,
Otis—Juvenile literature. [1. Underwater exploration. 2. Hydrothermal vents.
3. Beebe, William, 1877–1962. 4. Barton, Otis.] I. Title. II. Series: Matsen,
Bradford. Incredible deep-sea adventures.
GC65.M37 2003
551.46'07—dc21
                                                    2002013822

Printed in the United States of America

10 9 8 7 6 5 4 3 2 1

To Our Readers: We have done our best to make sure all Internet Addresses in this
book were active and appropriate when we went to press. However, the author and
the publisher have no control over and assume no liability for the material available
on those Internet sites or on other Web sites they may link to. Any comments or
suggestions can be sent by e-mail to comments@enslow.com or to the address on the
back cover.

Photo Credits: Robert Hartley/National Geographic Image Collection, p. 9; David
Knudsen/National Geographic Image Collection, pp. 10, 20; National Oceanic and
Atmospheric Administration, pp. 3, 5, 6, 13, 18, 19, 21, 23, 24–25, 27, 28, 30–31,
32, 33, 35, 36, 38, 40, 41; Charles E. Riddeford/National Geographic Society, p. 8;
© Wildlife Conservation Society, headquartered at the Bronx Zoo, pp. 4, 14, 26.

Cover Photos: National Oceanic and Atmospheric Administration; © Wildlife
Conservation Society, headquartered at the Bronx Zoo (Beebe, bathysphere).

Contents

1 Heroes of the Deep................................ 5

2 Voyage into the Depths.......................13

3 A Record Is Broken,
  A Record Is Set 27

4 Explorers of the Abyss 32

5 The Ultimate Dive to
  the Bottom of the Sea 37

Chapter Notes 43

Glossary 45

Further Reading 47

Internet Addresses................................47

Index 48

WILLIAM BEEBE AND THE BATHYSPHERE. IN THIS SMALL CRAFT, BEEBE AND OTIS BARTON WERE ABLE TO DIVE DEEPER THAN ANYONE HAD GONE BEFORE.

CHAPTER 1

Heroes
of
the Deep

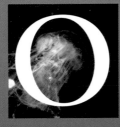**O**n a summer morning in 1934, William Beebe and Otis Barton prepared to go where no human had ever gone before. Their craft was not a beautiful space shuttle, but their expedition would be every bit as thrilling, important, and dangerous as a trip into space. They were going to dive a half mile down into the Atlantic Ocean in a hollow steel ball at the end of a cable.

They named their craft a bathysphere, which comes from Greek words that mean "sphere of the deep." In it, Beebe and Barton hoped they could set the world depth record. They also wanted to see the mysterious creatures of the deep ocean. The bathysphere weighed

5,000 pounds (2,268 kilograms), as much as two cars. They carried it to the site of their dive near the island of Bermuda on the deck of a barge named *Ready*. A ship named *Gladisfen* towed the *Ready*. Several tugboats helped, too.

The bathysphere would be lowered into the sea at the end of a thick cable by a powerful winch. The winch also had to be strong enough to pull the bathysphere back up again. The winch was on the deck of the *Ready*.[1]

The Brave Explorers

The men making the journey into the depths had to be very brave for two reasons. First, they were the first humans ever to make such a voyage so deep into the ocean. Second, their craft was a brand new invention. They tested their bathysphere many times, but every dive was risky. If the bathysphere was crushed or the cable broke, they would drown.

Both of the explorers were passionate about ocean adventures. They were also curious about the creatures of the deep sea. Dr. Will Beebe was a famous scientist who lived in New York City. He was best known for studying birds, but he loved the ocean, too.

Otis Barton was a wealthy, young student at Columbia University, also in New York City. He was very interested in the mysteries of the deep ocean. He was quite an inventor, too. Barton met Beebe, and they decided to explore the ocean depths together.

Beebe and Barton made many shallow dives in helmets and diving suits. Those dives, though, could go no deeper than sixty feet (eighteen meters) in the waters near the shore. In those days, diving with air tanks had not been invented. Divers were connected to a hose that ran up to an air pump on the surface.

The beautiful and strange creatures they saw on those early dives made them want to go deeper. Much deeper. In their bathysphere, they would try to dive more than fifty times deeper than any human could go in a diving suit. They would become true aquanauts. The word means "deep water voyager."

EARLY
DIVERS
COULD NOT
GO VERY
DEEP
BECAUSE
THEY WERE
ATTACHED
TO PUMPS
ON THE
SURFACE
BY A HOSE.

Building the Bathysphere

The bathysphere was a great invention. Otis Barton designed it, and the Watson Stillman Hydraulic Machinery Company in Roselle, New Jersey, built it. Barton paid the $12,000 cost of the bathysphere and its cables from his own pocket.[2]

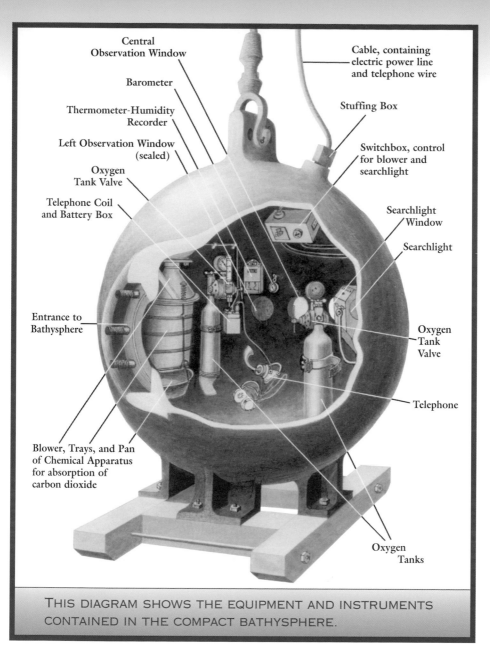

Central Observation Window

Barometer

Thermometer-Humidity Recorder

Left Observation Window (sealed)

Oxygen Tank Valve

Telephone Coil and Battery Box

Entrance to Bathysphere

Blower, Trays, and Pan of Chemical Apparatus for absorption of carbon dioxide

Cable, containing electric power line and telephone wire

Stuffing Box

Switchbox, control for blower and searchlight

Searchlight Window

Searchlight

Oxygen Tank Valve

Telephone

Oxygen Tanks

THIS DIAGRAM SHOWS THE EQUIPMENT AND INSTRUMENTS CONTAINED IN THE COMPACT BATHYSPHERE.

Otis Barton knew that the bathysphere had to be very strong to resist the enormous forces of the ocean depths. If the bathysphere was not strong enough, it would be crushed like a wad of tin foil. So, Barton designed the bathysphere in

THE MOTHER-
SHIP *READY*
CARRIED THE
BATHYSPHERE
TO THE SITE
OF BEEBE AND
BARTON'S
FAMOUS DIVE.

the shape of a perfectly round ball, or sphere. A sphere is the strongest shape for withstanding the pressure of the water in the deep ocean. Inside, the bathysphere was only 4.5 feet (1.4 meters) across, just big enough to hold two men sitting down or crouching. Its steel walls were one and a half inches thick. Beebe's nickname for the bathysphere was "the tank."

No matter how strong the bathysphere was, it could be lost if the cable or winch broke. Otis Barton ordered the cable from the famous Roebling brothers, who also made cables for great bridges. They made the cable for the well-known Brooklyn Bridge in New York City. The bathysphere cable was almost an inch thick and could lift and hold 29 tons (26,000 kilograms).

A rubber hose was also attached to the bathysphere to carry the wires for electricity for lights and a telephone line. This hose entered the craft through a hole on the top, next to

CLIMBING INTO THE BATHYSPHERE WAS NOT AN EASY FEAT!

the cable attachment. The hole was sealed tight with a special kind of connection called a stuffing box. One of Barton's jobs inside the bathysphere was to make sure the stuffing box did not leak.

Ready to Dive!

It was the day of the dive. The deep-sea aquanauts squeezed into the tiny cabin through a hole just fourteen inches wide, barely enough space to let them through. This hatch was covered by a thick, four-hundred-pound door and was sealed by ten heavy bolts.

Inside, Beebe and Barton crouched on the bare metal floor. They wore warm clothes because the water in the deep ocean is freezing cold. They could look out through two eight-inch portholes made of quartz glass three inches thick. A third porthole was sealed with a metal plate. They could turn on a single powerful spotlight with a switch inside the cabin.

In the bathysphere, Beebe and Barton breathed air from tanks inside the cabin. Exhaled air is full of a gas called carbon dioxide. Sealed in the bathysphere, Beebe and Barton had to have a way to get rid of that gas, because too much carbon dioxide can be poisonous. They had an open tray of special chemicals to take the carbon dioxide from the air they exhaled. The chemicals could only trap a certain amount of carbon dioxide, and then they would stop working. If Beebe and Barton stayed down too long, the carbon dioxide would kill them.[3]

Just like astronauts going into orbit, Beebe and Barton had a lot of help getting into their bathysphere and making their voyage to the bottom of the sea. Their support team had to be well trained for their special jobs. If any one of their crew made a mistake tightening the bolts on their hatch or tending the cable that lowered and raised the bathysphere, the explorers might never return alive.

In one of the earlier unmanned tests of the bathysphere, the craft was lowered a half mile into the sea to make sure its door and windows were tight enough. When the crew was bringing the craft up on its cable, they noticed that the winch was working very hard. The bathysphere seemed to be much heavier than usual. Once they had it on deck, a huge gush of water shot out from inside, and bolts were torn loose from the door. The bathysphere had leaked badly. If there had been people inside, they would have been crushed by the pressure of the water and drowned.

"Lower the Bathysphere"

Two members of the crew on the *Ready* were responsible for sealing the hatch. The hatch was sealed from the outside. Beebe and Barton would not be able to escape if an accident happened even at very shallow depth. Beebe and Barton knew they were making a very dangerous trip into the deep ocean.[4]

Another crew member named Gloria Hollister was in charge of communication to the bathysphere through the telephone wire. Her voice would be the only human contact Beebe and Barton would have as they descended into the dark depths of the sea. Another assistant, Jocelyn Crane, watched the marks on the winch cable. She kept track of the depth of the bathysphere. The attempt to set the record would be exciting, but very tense for the whole crew.[5]

After three years of test dives and preparation, Beebe and Barton were sealed into their bathysphere on the morning of August 15, 1934. They were ready for one of the greatest adventures of all time. Beebe gave the command to lower them into the ocean over the telephone to Gloria Hollister. She relayed the command to the winch operator—"Lower the Bathysphere."

Beebe and Barton were on their way.

CHAPTER 2

Voyage *into* the Depths

Down, down, down went Beebe and Barton. Around them, the bathysphere creaked and made frightening sounds. The pressure of the sea was pushing on the steel and rivets of the bathysphere. They had tested the hull down to a depth of 1,426 feet (435 meters), but this dive would be twice as deep. The water pressure would be much, much greater.

For each 33 feet (10 meters) they went down, the pressure on their craft doubled. This happened because the weight of the water pressing on them increased as they dove. Imagine piling rocks all around a beach ball.

William Beebe and Otis Barton stand on either side of their bathysphere.

Greater depth is like more rocks on the ball. That's how the water pressed on the bathysphere with Beebe and Barton inside.

Fortunately, the aquanauts were not in a beach ball. They were protected by the thickness of the steel and the shape of the bathysphere. They figured out that their steel ball would be crushed by the water pressure at about 4,500 feet (1,371 meters). Their goal was 3,000 feet (914 meters), or a little over a half mile down. If the cable broke and the bathysphere plunged much deeper than that, it and the men inside would be lost.

The Dive Begins

Inside the diving bathysphere, Beebe and Barton were too excited to worry about a breaking cable or other disaster. They looked through the two portholes and saw things never before seen by human eyes.

At first, they could see only the bubbles and foam from their splash into the sea. Then they could see the clear water in the beams of their two searchlights. Beebe and Barton peered anxiously through their tiny portholes. They immediately noticed that the colors were very different from those they were used to on land.

"At 9:41 in the morning we splashed beneath the surface," Beebe wrote, "and the sudden shift from a golden yellow world to a green one was unexpected. After the foam and bubbles passed from the glass, we were bathed in green; our faces, the tanks, the trays, even the blackened walls were tinged [with green]."[1]

The colors that we can see with our eyes range from red to violet, as in a rainbow. In between are orange, yellow, green, and blue. This is called the spectrum of colors. The aquanauts noticed that the familiar colors of the spectrum began to vanish as they went deeper. "The first plunge erases, to the eye, all the comforting warm rays of the spectrum. The red and the orange are as if they had never been, and soon the yellow is swallowed up in the green," Beebe wrote.[2]

Gradually, all the colors disappeared, and the sea in their portholes was dark, dark blue. Beebe switched on the spotlight. In its beam, he and Barton saw tiny particles drifting like dust in a ray of sunlight coming through a window at home. This ocean dust is made up of microscopic animals called plankton and tiny bits of other plants and animals floating in the water.

Studying the Creatures of the Sea

Before Barton invented the bathysphere, he and Beebe studied the animals of the deep ocean from the surface. Beebe dragged silk nets through the water and set traps to capture fish and other animals to study. This was not very successful. The creatures that came up in traps and nets were often crushed and deformed by their capture.

In the bathysphere, though, Beebe and Barton were the first human beings to see those creatures alive in their natural habitat. They had made several test dives, so they had already seen some of the creatures that lived down to 1,426 feet (435 meters). Even those shallower dives, though, had been thrilling to Beebe.

"After those dives were past, when I came again to examine the deep-sea treasures in my nets, I would feel as an astronomer might who looks through his telescope after having rocketed to Mars and back. Or like a paleontologist who could suddenly annihilate time and see his fossils alive," Beebe wrote after his test dives.[3]

Beebe and Barton liked the adventure of riding in a dangerous steel ball to set a new depth record. Even more, they liked seeing the creatures of the extreme depths down to 3,000 feet (914 meters). They were shocked by what they saw.

The Beauty of the Deep Sea

Beebe said that describing what he saw on his dive was one of the hardest things he ever did.[4] Fish and jellylike animals appeared in their searchlight and quickly vanished into the darkness of the sea. Some were curious and swam right up to the portholes and looked in at the two men.

About 300 feet (91 meters) down, Beebe saw amazing animals called siphonophores. These are individual jellylike creatures that live in huge colonies strung together like beads on a necklace. The biggest colonies can be 130 or more feet long. That's longer than a blue whale.[5] A colony of siphonophores brought up in a trap or net looked like a pile of clear, oozing glop. From the bathysphere, they appeared as beautiful as a string of jewels.[6]

"At 320 feet a lovely colony of siphonophores drifted past," wrote Beebe. "At this level they appeared like spun glass. . . . In our nets we find only the half-broken swimming bells, like cracked, crystal chalices, with all the wonderful loops

THESE JELLYLIKE SIPHONOPHORES LOOKED LIKE "A STRING OF JEWELS," SAID BEEBE.

and tendrils and animal flowers completely lost or contracted into a mass of tangled threads."[7] For Beebe and Barton, the greeting of the beautiful siphonophores was just the beginning of a wonderful show that no humans had ever seen before.

Almost everything the aquanauts would learn on their bathysphere dives would change the way people thought about the creatures of the sea. At 340 feet (104 meters), a foot-long pilot fish swam right up to the porthole and looked in.[8] Until Beebe saw the pilot fish at this depth, people thought they only lived near the surface, where they are often found swimming near sharks and turtles.

Around 400 feet (122 meters) down, Beebe saw big tuna called yellowfins. At about 500 feet (152 meters), slightly

LARGE TUNA WERE NOT AFRAID TO GET UP CLOSE TO THE BATHYSPHERE.

smaller fish called blue-banded jacks appeared. The fish were very curious about the bathysphere and swam right up to the windows, just like the pilot fish. While Beebe and Barton marveled at the visiting fish, silvery squid shot by the windows. The adventurers were surprised that the fish and squid did not seem to be afraid of the bathysphere or its lights.

All Is Well in the Bathysphere

Beebe and Barton were only one sixth of their way to their target depth. Already, they were making science history by observing the sea life around them. Inside the bathysphere, they were cramped in the small, dark space. The trip to 3,000 feet (914 meters) would take two hours, and the return to the surface an hour. Because of what they were seeing outside, though, the brave men did not think much about their discomfort.

GLORIA HOLLISTER, LEFT, AND OTHER CREW MEMBERS REMAINED IN CONSTANT CONTACT WITH BEEBE AND BARTON BY TELEPHONE.

Beebe and Barton took notes. Beebe described what he was seeing on the telephone to Gloria Hollister up on the deck of the *Ready*. She took notes, too. Beebe also carried a sketchbook with him to make quick drawings of the creatures he saw. Their accounts of the dive and what Beebe and Barton observed would be much more valuable to other scientists than just setting the depth record.

At 800 feet (244 meters), all was well in the bathysphere. The air tanks and carbon dioxide trays were working just fine. Every so often, the cable jerked up and down with the action of the waves on the surface, throwing the men around in the dark capsule. But they reported no injuries, so the mission to set the depth record was still a go.

"Never for a moment did either of us admit the possibility of failure," Beebe wrote about the risk of his dive. "Barton was sustained by his thorough knowledge of the mechanical margins of safety, and my hopes of seeing a new world of life left no opportunity for worry about possible defects."[9]

Living Light in the Deep Sea

As the bathysphere went down, the creatures outside became stranger. At about 1,600 feet (488 meters), all traces of sunlight faded from the water.[10] The water was pitch black except for the beam of the floodlight.

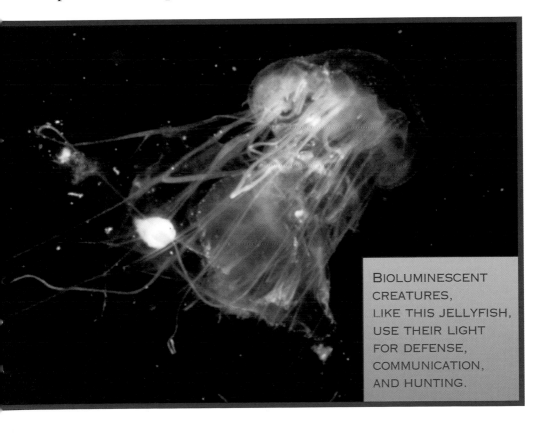

BIOLUMINESCENT CREATURES, LIKE THIS JELLYFISH, USE THEIR LIGHT FOR DEFENSE, COMMUNICATION, AND HUNTING.

At great depths where there is no sunlight, fish and other creatures have developed bioluminescence. The word means "living light." On land, you can see bioluminescence in fireflies. In the sea, many fish and jellyfish give off their own light.

Usually, a fish will use its lights when it is frightened or hunting for a meal. A fish seeing the bathysphere and its searchlight for the first time would probably be both scared and ready to do battle if necessary. Therefore, Beebe and Barton saw a constant light show as they descended into the depths.

Just past 2,000 feet (610 meters), Beebe spotted a bioluminescent fish that was bigger than the bathysphere. It looked a lot like big fish he had seen near the surface, but with much bigger eyes and a shorter jaw. He saw dots of pale blue lights along its body. He also saw two tentacles. One had a red light at the end, and the other had a blue light. They were glowing lures to attract the fish's prey! They twitched and jerked like fishing lures used in ponds and streams.

Beebe named this incredible fish a sea dragon. It held its mouth open all the time, showing big, sharp fangs, which were lit by more lights from inside its throat. These lights in its throat attracted other fish to swim right into the monster's mouth. Then . . . CHOMP![11]

Monster of the Deep

As Beebe and Barton neared their record depth, they saw a fish that was even more shocking than the sea dragon. A huge shape appeared in the dim light cast by the bathysphere's searchlight. As the aquanauts watched in horror, this giant of the deep swam past at the very edge of the light. They estimated it to be at least twenty feet (six meters) long, but probably much longer. They saw no fins or eyes, only a wall of flesh sliding by them.

"Whatever it was, it appeared and vanished so unexpectedly and showed so dimly that it was quite unidentifiable except as a large, living creature," Beebe wrote. He and Barton were "slightly shaken" by the encounter. When Beebe thought later about this encounter with the giant creature, he recalled old reports about sea serpents.[12]

And down they went. Monsters never seen before were swimming around them. The water pressure was trying to crush them. The bathysphere shuddered at the end of its cable. Beebe and Barton called on all their courage to continue to their goal of a half mile down.

BEEBE AND BARTON SAW MANY UNUSUAL AND SCARY-LOOKING FISH, LIKE THIS SCORPIONFISH.

The *Incredible* Bathysphere[13]

COST
$12,000

WEIGHT
5,000 pounds (2,268 kilograms)

MADE OF
Steel

CABIN SIZE
4½ feet (1½ meters) in diameter

THICKNESS OF WALLS
1½ inches (4 centimeters)

SIZE OF CABLES
⅞ inch (2 centimeters)

STRENGTH OF CABLE
29 tons (26,000 kilograms)

FIRST UNMANNED TESTS
June 3, 1930

RECORD DIVE
3,028 feet (923 meters);
August 15, 1934

More *Fun* Facts

William Beebe called the bathysphere "the tank."
He described it as "rather like an enormous inflated
and slightly cockeyed bullfrog."

Otis Barton would not make a dive without
his lucky hat. One dive was held up so he could find
his hat. The entire crew ran around the *Ready*
and the *Gladisfen* trying to find it.

Beebe did not like machines. He hated to drive
a car. Beebe gave his crew rides in the
bathysphere on their birthdays.

Beebe and Barton tied bait to the outside of
their portholes to attract fish to the bathysphere.

In 1933, the bathysphere
was exhibited at the
World's Fair in Chicago.

A movie about the bathysphere
was made in 1937.
It was called *Titans of the Deep.*

Beebe looks out of the bathysphere's hatch.

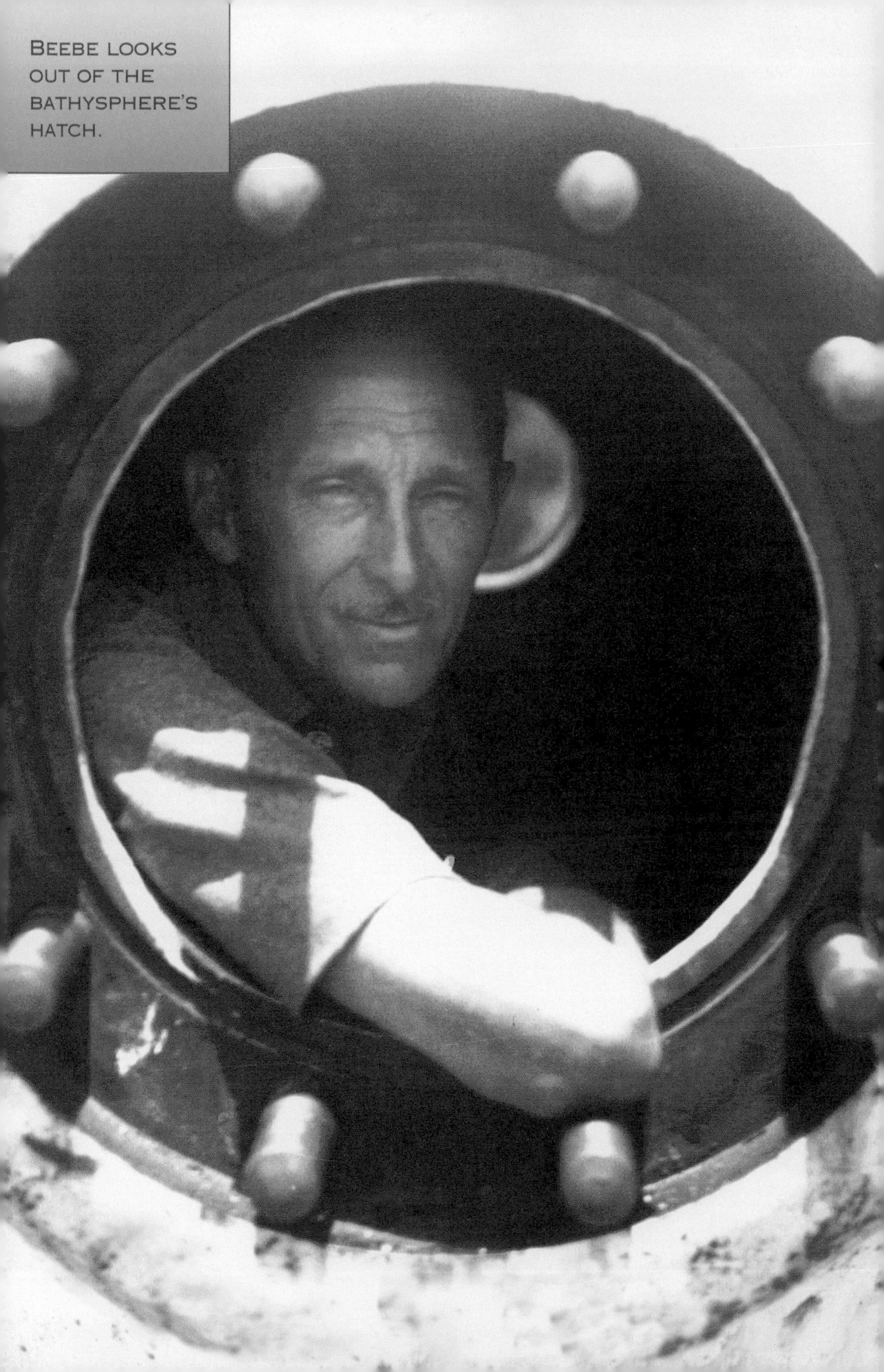

CHAPTER 3

A Record Is *Broken,* A Record Is Set

illiam Beebe and Otis Barton broke their old record for diving beneath the sea when the bathysphere reached 1,427 feet (435 meters). (They had already made a test dive to 1,426 feet.) As they passed through the new record depth, Gloria Hollister held her telephone in the direction of some tugboats that were helping with the expedition. The tugboats blew their whistles and Barton and Beebe heard the whistles over a quarter of a mile down in the sea through their telephone.[1]

Beebe, Barton, and other experts had estimated how deep the bathysphere could dive without being crushed

THE TWO EXPLORERS SAW AN ENDLESS PARADE OF UNIQUE SEA LIFE, SUCH AS THIS BATFISH.

but no one really knew for sure. The courageous Beebe ordered the crew of the *Ready* to continue lowering the bathysphere. Finally, just past a half mile (2,640 feet), the new record was set at 3,028 feet (923 meters). This record would stand for fourteen years.[2]

A Half Mile Down

Beebe and Barton hung at the end of the cable at 3,028 feet. They were cold and uncomfortable, but alive where no human had ever been alive before them. Like future astronauts on the moon, they knew they were

pioneers. They continued to observe and make notes about what they could see out their portholes.

A half mile down, Beebe sat with his forehead pressed against the glass of the porthole. He covered his mouth and nose with a handkerchief so the window would not fog up. Beebe was a brave and careful explorer, but he was also a human being. He became very emotional as he watched the creatures of the depths pass his window.

"There came to me at that instant a tremendous wave of emotion, a real appreciation of what was momentarily almost superhuman, cosmic, of the whole situation," Beebe wrote later. "We dangled in mid-water, isolated as a lost planet in outermost space."[3]

A Scary Noise

After only five minutes at 3,028 feet, Captain Sylvester of the mother ship *Gladisfen* called down and told Beebe that he must begin the trip to the surface immediately. The captain was afraid that the winch might have trouble bringing them back from such a depth.

Beebe agreed with the captain. "Over two hours had passed since we left the deck and I knew that the nerves of both my staff and myself were getting ragged with constant tenseness and strain. So I asked for our ascent."[4]

"We are prepared to ascend," Beebe said to Gloria Hollister over the telephone line to the bathysphere. On the deck of the *Ready*, the crew sprang into action. The winch began to turn. Deep down in the sea, the men in the bathysphere heard a loud plunk.

"What happened? What happened?" Beebe yelled anxiously over the telephone. The bathysphere was bobbing up and down, far beyond any hope of rescue if the cable or the winch broke. The voices from the *Ready* up above were also anxious. A rope that guided the steel cable onto the reel of the winch had torn off. That was what made the loud plunk.[5]

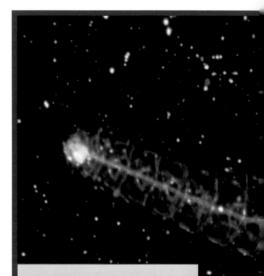

Beebe and Barton and the rest of the crew relaxed. The trip to the surface began. As they rose through the water, the bathysphere lurched and swayed. The aquanauts continued to take notes on what they could see out the window. Fish, jellyfish, and creatures no one could identify flashed in and out of their spotlight.

DEEP UNDER THE OCEAN'S SURFACE, IT LOOKED AS IF BEEBE AND BARTON WERE IN AN ENTIRELY DIFFERENT WORLD. CREATURES SUCH AS THESE SIPHONOPHORES WERE SEEN FOR THE FIRST TIME IN THE DEEP OCEAN.

Finally, just after noon, the bathysphere bobbed up next to the *Ready*. Everybody on the barge breathed a sigh of relief. The winch lifted the bathysphere to the deck. The crew could see Beebe and Barton through the portholes. They were smiling. Then the hatch was opened, and the great adventure ended with a wild celebration on the deck of the *Ready*.

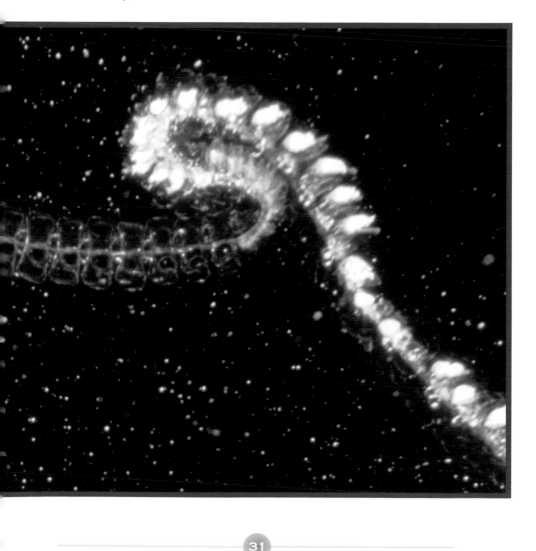

CHAPTER 4

Explorers
of the
Abyss

Setting the new record made Beebe and Barton celebrities. They were as famous in 1934 as Neil Armstrong and Buzz Aldrin were in 1969 after they became the first humans to walk on the moon. They were as famous as rock stars are today. They were special heroes to children, and teachers told stories of their dives in school classes. Beebe and Barton wrote magazine articles and books. Both knew that they had done something very heroic. Beebe thought the deep ocean was as mysterious and interesting as outer space.[1]

Beebe and Barton did not risk their lives just for the adventure of diving so deep into the sea. They knew that their example would inspire other explorers. They knew that their record would challenge others to break it. And they knew that their descriptions of what they saw would be precious to other scientists.

Treasures from the Deep

The notes and drawings made by Beebe and Barton were treasures from the deep for scientists of all kinds. The two explorers made hundreds of speeches about what they had done and seen. In all their books, articles, and speeches, they pointed out how much there was left unknown about the sea. Because scientists love questions as much as answers, the sea gives them wonderful challenges. Every discovery brings up more questions.

BEEBE AND BARTON'S DISCOVERIES HAVE HELPED SCIENTISTS DISCOVER MORE ABOUT MARINE LIFE. THIS TILEFISH LIVES DOWN TO HUNDREDS OF FEET BELOW THE SURFACE.

Human beings have still only seen a small number of the creatures of the sea. All those strange fish and jellyfish that Beebe and Barton saw were just the beginning. In the sixty years since the descent off Bermuda, hundreds and hundreds of new kinds of animals have been discovered in the ocean. Scientists know, though, that thousands more remain to be found.

Inspiring New Explorers

Beebe and Barton made many more dives in the bathysphere. They also taught other adventurers the secrets of deep diving. Their courage was very inspiring to ocean scientists and explorers. Many say that the two heroes of the tiny bathysphere and its exciting dives are the reason they chose to study the oceans.

Dr. Cindy Lee Van Dover is one of these. She is now a well-known ocean scientist. She was also the first woman pilot of the deep submersible named *Alvin*. She says she would not have become an oceanographer without the inspiration of Will Beebe and Otis Barton.[2]

Another ocean explorer who was inspired by William Beebe's courage and discoveries is Dr. Sylvia Earle. She became a famous oceanographer and, for a while, was the chief ocean scientist of the United States. "I fell in love with the stories by William Beebe," Earle said.[3]

"Beebe discussed what it was like to go down inside a submersible and peer out of a porthole and see beautiful, luminescent fish, with lights down the side like ocean liners," Earle said, "bizarre creatures of the sort that you just don't

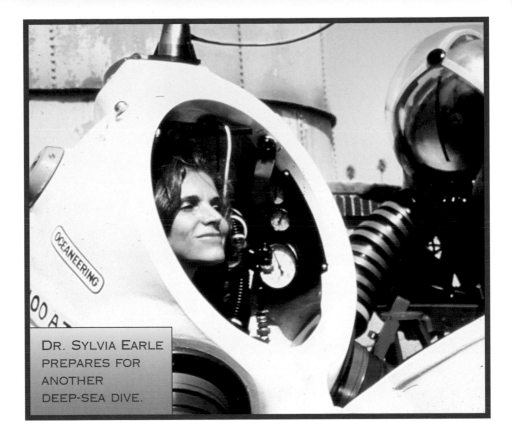

DR. SYLVIA EARLE PREPARES FOR ANOTHER DEEP-SEA DIVE.

see walking down the street, or going into the forest, or even looking around in shallow water. The aquariums of the world, as wonderful and diverse as they are, do not have the sort of creatures that Beebe described from his exploration back in the 1930s. I found that utterly inspiring."[4]

Cindy Lee Van Dover, Sylvia Earle, and thousands of other men and women have become deep-ocean explorers since Beebe and Barton led the way in their tiny bathysphere. And we have invented new kinds of craft to travel even deeper into the abyss.

CHAPTER 5

The Ultimate Dive *to the* Bottom of the Sea

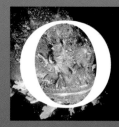

Otis Barton, William Beebe, and all the people who were inspired by them continued the exploration of the abyss. The next step was a craft that needed no winch or cable to go up and down in the ocean. It was called a bathyscaph. The name is from Greek words that mean "deep boat." The bathyscaph was invented by another hero of the deep named Auguste Piccard.

Auguste and Jacques Piccard

Auguste Piccard had already invented a high-altitude balloon with a cabin to carry a pilot.

He rode in it up to a height of nine miles in 1931. He was inspired by adventures of Beebe and Barton in the ocean and set to work on a new kind of craft for diving into the abyss.[1] Ten years later, he took his new bathyscaph for a dive.

The bathyscaph was a giant step in ocean exploring. It had no cable, so it could dive all the way to the bottom of the sea and land there. Its crew cabin was seven feet wide, and its walls were much thicker than those of the bathysphere. This meant that the bathyscaph could dive deeper than the bathysphere.

The new craft was shaped more like a football than a beach ball. It also had a big tank filled with gasoline above the crew cabin. Gasoline is lighter than seawater, so the bathyscaph would float. To submerge, the pilot opened two other tanks with a remote control. Seawater flowed into those tanks, and down went the bathyscaph. To rise from the bottom, the pilot released iron pellets from the belly of his craft, making it lighter.[2]

The first bathyscaph could not move around very much in the water. It could only go up and down. By 1953, Piccard had invented an improved bathyscaph that could cruise around, as well as go up and down. He named it *Trieste*, after a city in Italy.

In *Trieste*, Piccard and his son Jacques dove straight down 2 miles (3.2 kilometers) to the bottom of the sea near Naples, Italy. They landed with a bump. When they came back to the surface, the bathyscaph was covered with the mud of the seafloor. The *Trieste* aquanauts were disappointed on their first dive. The bottom off Naples was not very interesting. There were few creatures and the seafloor was just a flat muddy plain.[3]

Trieste and the Challenger Deep

The best was yet to come, though. After Piccard proved that the *Trieste* worked, the United States Navy bought his bathyscaph and hired him to be its pilot. Piccard also trained other men to pilot *Trieste*. He and his crews made many dives to the bottom all over the ocean. Like Beebe and Barton, the aquanauts of *Trieste* kept notes about what they saw in the depths. They also made maps of the seafloor at many different depths.

Finally, in 1960, Jacques Piccard and a submarine officer, Navy Lieutenant Don Walsh, climbed into *Trieste* to dive to the deepest spot in the ocean. They were near the island of Guam in the Pacific Ocean. There, the bottom of the sea was about seven miles (eleven kilometers) beneath them. The place is named the Challenger Deep after a famous British

THE BATHYSCAPH *TRIESTE* DOVE ALMOST FOURTEEN TIMES AS DEEP AS THE BATHYSPHERE.

ship. Challenger Deep is part of a huge canyon on the sea bottom named the Marianas Trench.

On the morning of January 23, 1960, the surface of the sea was rough, and the sky above was stormy. Just twenty-five years after Beebe and Barton dove a half mile down, Piccard and Walsh were trying for seven miles! The pressure at that depth was enormous. If something went wrong, there was no hope for a rescue.

The Dive to the Bottom of the Sea

Piccard gave the command to dive. The brave explorers plunged down into the darkness of the abyss. Their voyage down and back to the surface would take nine hours. Down through the depths they went. The water became pitch black. Strange creatures flashed and swam around the bathyscaph. Inside, Piccard and Walsh were in awe of what they saw.

As they neared the deepest place on earth, *Trieste* shook with a tremendous explosion. Don Walsh calmly asked Piccard if they had landed on the bottom. "I do not believe so," Piccard replied. He figured out that one of *Trieste*'s spotlights had exploded and caused the noise.

A few minutes later, they saw the seafloor. In their lights, it looked like a wilderness of ooze and mud. Then, two amazing things happened that shocked Piccard and Walsh. As they watched the bottom, a foot-long flatfish wriggled out of the mud and swam away. Then, a tiny red shrimp swam by the window. Even at the greatest depth, there is life in the sea![4]

China

India

Philippines

Philippine Trench

Mariana Trench

Challenger Deep

Indonesia

AUSTRALIA

TRIESTE DOVE DOWN TO A HUGE CANYON CALLED CHALLENGER DEEP ON THE BOTTOM OF THE PACIFIC OCEAN.

The dive of the *Trieste* to the very bottom of the sea in the Challenger Deep was big news. The heroism of Piccard and Walsh and the success of the bathyscaph encouraged a boom of deep-ocean discovery. Using Piccard's ideas, other explorers began building bathyscaphs. Soon, *Alvin*, *Halibut*, *Sea Cliff*, and other deep-diving craft were also exploring the abyss.

The Adventure Continues

We now know that fish and other animals live everywhere in the sea, even in the deepest parts. We have maps of what most of the floor of the sea looks like. We know that the sea is both very powerful and very

THE SUBMERSIBLE *ALVIN* ENABLED SCIENTISTS TO DISCOVER AN ENTIRELY NEW TYPE OF UNDERWATER LIFE.

fragile. We know that we can hurt the sea and its plants and animals if we pollute it with chemicals, plastics, and oil.

Many questions about the deep ocean and its creatures have been answered by heroes of the deep such as William Beebe, Otis Barton, Jacques Piccard, Don Walsh, Cindy Lee Van Dover, and Sylvia Earle. Many questions remain to be answered. The exploration will continue as long as humans are brave and curious. Perhaps you will be part of this great adventure.

Chapter Notes

CHAPTER 1. HEROES OF THE DEEP

1. Catherine L. Hines, *The Official William Beebe Web Site*, "The Bathysphere Adventures," © 2000, <http://hometown.aol.com/chines6930/mwl/sphere.htm> (July 1, 2002).

2. Ibid.

3. William Broad, *The Universe Below* (New York: Simon and Schuster, 1997), pp. 39–41.

4. Ibid.

5. Hines.

CHAPTER 2. THE VOYAGE INTO THE DEPTHS

1. William Beebe, *Half Mile Down* (New York: Duell Sloan Pearce, 1951), n.d., <http://seawifs.gsfc.nasa.gov/OCEAN_PLANET/HTML/ocean_planet_book_beebe1. html>
(July 1, 2002).

2. Ibid.

3. William Beebe, *Adventuring with Beebe* (New York: The Viking Press, 1951), p. 84.

4. Ibid.

5. William Broad, *The Universe Below* (New York: Simon and Schuster, 1997), p. 204.

6. Beebe, *Half Mile Down*.

7. Ibid.

8. Ibid.

9. William Beebe, "A Roundtrip to Davy Jones's Locker," *National Geographic*, June 1931, p. 655.

10. Judith Gradwohl, ed., *Ocean Planet* (New York: Smithsonian and Harry N. Abrams, 1995), p. 99.

11. William Broad, *The Universe Below* (New York: Simon and Schuster, 1997), pp. 41–42.

12. Ibid.

13. Catherine L. Hines, *The Official William Beebe Web Site*, "The Bathysphere Adventures," © 2000, <http://hometown.aol.com/chines6930/mwl/sphere.htm> (July 1, 2002).

Chapter Notes

CHAPTER 3. A RECORD IS BROKEN, A RECORD IS SET

1. Judith Gradwohl, ed., *Ocean Planet* (New York: Smithsonian and Harry N. Abrams, 1995), p. 96.

2. Ibid., p. 99.

3. Catherine L. Hines, *The Official William Beebe Web Site*, "The Bathysphere Adventures," © 2000, <http://hometown.aol.com/chines6930/mwl/sphere.htm> (July 1, 2002).

4. William Beebe, *Half Mile Down* (New York: Duell Sloan Pearce, 1951), n.d., <http://seawifs.gsfc.nasa.gov/OCEAN_ PLANET /HTML/ocean_planet_book_beebe1.html> (February 23, 2002).

5. Hines.

CHAPTER 4. EXPLORERS OF THE ABYSS

1. *National Geographic*, "A Half Mile Down," December 1934, p. 704.

2. Cindy Lee Van Dover, *The Octopus's Garden: Hydrothermal Vents and Other Mysteries of the Deep Sea* (New York: Addison-Wesley Publishing, 1996), pp. 6–7.

3. Catherine L. Hines, *The Official William Beebe Web Site*, "The Bathysphere Adventures," © 2000, <http://hometown.aol.com/chines6930/mwl/sphere.htm> (February 23, 2002).

4. Ibid.

CHAPTER 5. THE ULTIMATE DIVE TO THE BOTTOM OF THE SEA

1. William Broad, *The Universe Below* (New York: Simon and Schuster, 1997), p. 49.

2. Kathy Svitil, *PBS Online*, "Journey to the Ocean Floor," n.d., <http://www.pbs.org/wnet/savageseas/deep-side journey.html> (July 1, 2002).

3. Broad, p. 51.

4. Ibid., pp. 54–55.

Glossary

abyss—The very deep parts of the ocean.

aquanaut—A deep-water voyager. (*Astronaut* means "star voyager.")

astronomer—A scientist who studies the stars and planets.

bathyscaph—A manned research submarine that can operate without a cable because its pilots can control its buoyancy.

bathysphere—A submersible steel chamber that is lowered into deep water by a cable.

bioluminescence—Light produced by a living thing such as a fish or jellyfish.

carbon dioxide—A gas that is exhaled after a human or other animal breathes in air.

diving suit—A suit made of rubber with a heavy metal helmet that normally allows a person to dive no deeper than sixty feet using air pumped from the surface through a hose.

oceanographer—A scientist who studies the ocean itself, along with the fish and other creatures.

paleontologist—A scientist who studies life in the past by looking at fossils.

quartz glass—Glass made from very pure sand that has been melted down and formed into windows.

Glossary

siphonophore—A jellylike animal related to jellyfish that lives with other siphonophores in enormous chains called colonies.

spectrum—The colors visible to the eye.

stuffing box—A device that prevents leakage where a cable passes through a hole in a bathysphere or other vessel.

submersible—A craft for diving beneath the surface of the sea.

winch—A machine for lowering or raising an object at the end of a cable.

Further Reading

Dipper, Frances. *Mysteries of the Ocean Deep*. Brookfield, Conn.: Copper Beech Books, 1996.

Earle, Sylvia A. *Dive!: My Adventures in the Deep Frontier*. Washington, D.C.: National Geographic Society, 1998.

Earle, Sylvia A. *Sea Critters*. Washington, D.C.: National Geographic Society, 2000.

Kovacs, Deborah. *Dive to the Deep Ocean: Voyages of Ocean Exploration and Discovery*. Austin, Tex.: Raintree-Steck Vaughn Publishers, 2001.

Lalley, Pat. *Ocean Scientists*. Austin, Tex.: Raintree-Steck Vaughn Publishers, 2001.

Markle, Sandra. *Pioneering Ocean Depths*. New York: Atheneum Books for Young Readers, 1995.

Internet Addresses

Hines, Catherine. *The Official William Beebe Web Site*. "The Bathysphere Adventures." © 2002. <http://hometown.aol.com/chines6930/mw1/beebe1.htm>

Public Broadcasting System. *PBS Online*. "Savage Seas.", http://www.pbs.org/wnet/savageseas/deep-main.html>

U.S. Navy. *Office of Naval Research, Science and Technology Focus Site*. "People Under the Sea: Submersibles 1900–1960." <http://www.onr.navy.mil/focus/ocean/vessels/submersibles6.htm>

Index

A
Alvin, 34, 40, 41

B
Barton, Otis
  background, 7
  bathysphere expedition, 5,
    13–23, 30–31
  breaks record, 27–28
  designs bathysphere, 7–9
  fame, 32–33
  influences on others, 34–35,
    36, 37
  preparation for dive, 7, 10–12
batfish, 28
bathyscaph, 36–37
bathysphere
  cable, 5–6, 9, 11, 12, 15, 20,
    24, 28, 30
  construction of, 7–10
  cost, 7
  diagram, 8
  hatch, 10, 12, 31
  origin of name, 5
  portholes, 11, 15–16, 17–18,
    26, 29, 31
  size, 5–6
Beebe, William
  background, 7
  bathysphere expedition, 5,
    13–23, 29–31
  breaks record, 27–28
  fame, 32–33
  influences on others, 34–35,
    36, 37
  preparation for dive, 7, 10–12
Bermuda, 6, 34
bioluminescence, 21–22
blue-banded jacks, 19

C
carbon dioxide, 11
Challenger Deep, 38–40
Crane, Jocelyn, 12

E
Earle, Sylvia, 34–35, 42

G
Gladisfen, 6, 25, 29

H
Hollister, Gloria, 12, 20,
  27, 30

M
Marianas Trench, 39

P
Piccard, Auguste 36–37
Piccard, Jacques, 37–40, 42
pilot fish, 18

R
Ready, 6, 9, 20, 25, 28, 30–31
Roebling Brothers, 9

S
scorpionfish, 23
sea dragon, 22
siphonophores, 17–18, 30–31
Sylvester, Captain, 29

T
tilefish, 33
Trieste, 37–40
tuna, 18–19

V
Van Dover, Cindy Lee, 34,
  35, 42

W
Walsh, Don, 38–39, 42
Watson Stillman Hydraulic
  Machinery Company, 7

2/03